ENERGY SECTOR STANDARD OF THE PEOPLE'S REPUBLIC OF CHINA

中华人民共和国能源行业标准

Code for Design of Roller-Compacted Concrete Arch Dams

碾压混凝土拱坝设计规范

NB/T 10335-2019

Chief Development Department: China Renewable Energy Engineering Institute
Approval Department: National Energy Administration of the People's Republic of China
Implementation Date: June 1, 2020

China Water & Power Press

中国水利水电出版社

Beijing 2024

All rights reserved. No part of this publication may be reproduced, stored in a retrieval system, or transmitted in any form or by any means—electronic, mechanical, photocopying, recording or otherwise, without prior written permission of the publisher.

图书在版编目（CIP）数据

碾压混凝土拱坝设计规范：NB/T 10335-2019 = Code for Design of Roller-Compacted Concrete Arch Dams (NB/T 10335-2019)：英文 / 国家能源局发布. 北京：中国水利水电出版社，2024.9. -- ISBN 978-7-5226-2739-7

Ⅰ. TV541-65

中国国家版本馆CIP数据核字第202448DL32号

ENERGY SECTOR STANDARD
OF THE PEOPLE'S REPUBLIC OF CHINA
中华人民共和国能源行业标准

Code for Design of Roller-Compacted Concrete Arch Dams
碾压混凝土拱坝设计规范
NB/T 10335-2019
（英文版）

Issued by National Energy Administration of the People's Republic of China
国家能源局　发布
Translation organized by China Renewable Energy Engineering Institute
水电水利规划设计总院　组织翻译
Published by China Water & Power Press
中国水利水电出版社　出版发行
　　Tel: (+ 86 10) 68545888　68545874
　　sales@mwr.gov.cn
　　Account name: China Water & Power Press
　　Address: No.1, Yuyuantan Nanlu, Haidian District, Beijing 100038, China
　　http://www.waterpub.com.cn
中国水利水电出版社微机排版中心　排版
北京中献拓方科技发展有限公司　印刷
184mm×260mm　16开本　2.5印张　79千字
2024年9月第1版　2024年9月第1次印刷

Price（定价）：￥400.00

Introduction

This English version is one of China's energy sector standard series in English. Its translation was organized by China Renewable Energy Engineering Institute authorized by National Energy Administration of the People's Republic of China in compliance with relevant procedures and stipulations. This English version was issued by National Energy Administration of the People's Republic of China in Announcement [2023] No. 1 dated February 6, 2023.

This version was translated from the Chinese Standard NB/T 10335-2019, *Code for Design of Roller-Compacted Concrete Arch Dams*, published by China Water & Power Press. The copyright is reserved by National Energy Administration of the People's Republic of China. In the event of any discrepancy in the implementation, the Chinese version shall prevail.

Many thanks go to the staff from the relevant standard development organizations and those who have provided generous assistance in the translation and review process.

For further improvement of the English version, any comments and suggestions are welcome and should be addressed to:

China Renewable Energy Engineering Institute
No. 2 Beixiaojie, Liupukang, Xicheng District, Beijing 100120, China
Website: www.creei.cn

Translating organizations:

POWERCHINA Guiyang Engineering Corporation Limited

China Renewable Energy Engineering Institute

Translating staff:

HAO Peng	QIU Huanfeng	MIAO Jun	ZHANG Heng
LONG Qihuang	WU Yingang	YANG Chunping	CUI Jin
YANG Taoping	CHEN Tuo		

Review panel members:

JIN Feng	Tsinghua University
LIU Xiaofen	POWERCHINA Zhongnan Engineering Corporation Limited
ZHENG Xing	POWERCHINA Guiyang Engineering Corporation Limited

QIE Chunsheng	Senior English Translator
HUANG Zhibin	Guizhou Wujiang Hydropower Development Co., Ltd.
ZHOU Yuefei	POWERCHINA Zhongnan Engineering Corporation Limited
QIAO Peng	POWERCHINA Northwest Engineering Corporation Limited
JIA Haibo	POWERCHINA Kunming Engineering Corporation Limited
ZHANG Ming	Tsinghua University
YAN Wenjun	Army Academy of Armored Forces, PLA
LI Shisheng	China Renewable Energy Engineering Institute

National Energy Administration of the People's Republic of China

翻译出版说明

本译本为国家能源局委托水电水利规划设计总院按照有关程序和规定，统一组织翻译的能源行业标准英文版系列译本之一。2023年2月6日，国家能源局以2023年第1号公告予以公布。

本译本是根据中国水利水电出版社出版的《碾压混凝土拱坝设计规范》NB/T 10335—2019 翻译的，著作权归国家能源局所有。在使用过程中，如出现异议，以中文版为准。

本译本在翻译和审核过程中，本标准编制单位及编制组有关成员给予了积极协助。

为不断提高本译本的质量，欢迎使用者提出意见和建议，并反馈给水电水利规划设计总院。

地址：北京市西城区六铺炕北小街2号
邮编：100120
网址：www.creei.cn

本译本翻译单位：中国电建集团贵阳勘测设计研究院有限公司
　　　　　　　　水电水利规划设计总院
本译本翻译人员：郝　鹏　邱焕峰　苗　君　张　恒
　　　　　　　　龙起煌　吴银刚　杨春平　崔　进
　　　　　　　　杨桃萍　陈　拓
本译本审核人员：
　　金　峰　清华大学
　　刘小芬　中国电建集团中南勘测设计研究院有限公司
　　郑　星　中国电建集团贵阳勘测设计研究院有限公司
　　郄春生　英语高级翻译
　　黄志斌　贵州乌江水电开发公司
　　周跃飞　中国电建集团中南勘测设计研究院有限公司
　　乔　鹏　中国电建集团西北勘测设计研究院有限公司
　　贾海波　中国电建集团昆明勘测设计研究院有限公司
　　张　明　清华大学

闫文军　中国人民解放军陆军装甲兵学院
李仕胜　水电水利规划设计总院

国家能源局

Announcement of National Energy Administration of the People's Republic of China [2019] No. 8

National Energy Administration of the People's Republic of China has approved and issued 152 energy sector standards including *Code for Operating and Overhauling of Excitation System of Small Hydropower Units* (Attachment 1), and the English version of 39 energy sector standards including *Code for Safe and Civilized Construction of Onshore Wind Power Projects* (Attachment 2).

Attachments: 1. Directory of Sector Standards
2. Directory of English Version of Sector Standards

National Energy Administration of the People's Republic of China

December 30, 2019

Attachment 1:

Directory of Sector Standards

Serial number	Standard No.	Title	Replaced standard No.	Adopted international standard No.	Approval date	Implementation date
...						
10	NB/T 10335-2019	Code for Design of Roller-Compacted Concrete Arch Dams			2019-12-30	2020-07-01
...						

Foreword

According to the requirements of Document GNKJ [2014] No. 298 issued by National Energy Administration of the People's Republic of China, "Notice on Releasing the Development and Revision Plan of the First Batch of Energy Sector Standards in 2014", and after extensive investigation and research, summarization of practical experience, consultation of relevant advanced standards at home and abroad and wide solicitation of opinions, the drafting group has prepared this code.

The main technical contents of this code include: basic requirements, layout of RCC arch dam, RCC of dam, stress and stability analysis of arch dam, dam detailing, crack prevention and temperature control, safety monitoring design, construction requirements, and initial impoundment, operation and maintenance.

National Energy Administration of the People's Republic of China is in charge of the administration of this code. China Renewable Energy Engineering Institute has proposed this code and is responsible for its routine management. Energy Sector Standardization Technical Committee on Hydropower Investigation and Design (NEA/TC15) is responsible for the explanation of the specific technical contents. Comments and suggestions in the implementation of this code should be addressed to:

China Renewable Energy Engineering Institute
No. 2 Beixiaojie, Liupukang, Xicheng District, Beijing 100120, China

Chief development organizations:

POWERCHINA Guiyang Engineering Corporation Limited

China Renewable Energy Engineering Institute

Participating development organizations:

POWERCHINA Chengdu Engineering Corporation Limited

POWERCHINA Zhongnan Engineering Corporation Limited

POWERCHINA Northwest Engineering Corporation Limited

POWERCHINA Beijing Engineering Corporation Limited

POWERCHINA Huadong Engineering Corporation Limited

POWERCHINA Kunming Engineering Corporation Limited

China Institute of Water Resources and Hydropower Research

Chief drafting staff:

FAN Fuping	YANG Jiaxiu	WEI Xiaoming	CUI Jin
LUO Hongbo	XU Lin	TAN Jianjun	GUO Yong
LI Yunliang	FANG Guangda	CHEN Nengping	LONG Qihuang
YIN Huaan	XIAO Feng	CHEN Yongfu	DENG Yiguo
SU Yan	YE Jianqun	YU Jianqing	XIANG Hong
YANG Bo	HAO Peng	CHEN Yifeng	QIU Huanfeng
LIU Rongli			

Review panel members:

DANG Lincai	JIN Feng	FANG Kunhe	LIU Yi
WANG Qingyuan	WANG Yiming	CHEN Xiangrong	XU Jianqiang
WENG Yin	ZHOU Yuefei	CHEN Qiuhua	SHAO Jingdong
ZHANG Jing	WANG Guojin	LI Linian	ZHAO Yi
PANG Mingliang	LI Shuangbao	ZHANG Xiong	WANG Fuqiang
DU Xiaokai	LI Shisheng		

Contents

1	**General Provisions**	1
2	**Terms**	2
3	**Basic Requirements**	3
4	**Layout of RCC Arch Dam**	4
4.1	General Requirements	4
4.2	Dam Shape Design	4
4.3	Structures in Dam Body	5
5	**RCC of Dam**	7
5.1	General Requirements	7
5.2	Strength Class of RCC	7
5.3	Physical and Mechanical Properties of RCC	8
5.4	Seepage Resistance and Durability Performance of RCC	8
6	**Stress and Stability Analysis of Arch Dams**	10
7	**Dam Detailing**	11
7.1	Dam Crest Layout	11
7.2	Zoning Design	11
7.3	Contraction Joints	12
7.4	Joint Grouting	13
7.5	Galleries and Adits	14
7.6	Waterstops and Drains	15
8	**Crack Prevention and Temperature Control**	16
8.1	General Requirements	16
8.2	Temperature Control Standard	17
8.3	Temperature Control Measures	18
9	**Safety Monitoring Design**	19
9.1	General Requirements	19
9.2	Monitoring Items and Arrangement	19
9.3	Requirements for Instrument Installation	20
10	**Construction Requirements**	21
11	**Initial Impoundment, Operation and Maintenance**	23
Explanation of Wording in This Code		25
List of Quoted Standards		26

1 General Provisions

1.0.1 This code is formulated with a view to standardizing the design of roller-compacted concrete (RCC) arch dams and ensuring the design quality to achieve the objectives of safety and reliability, economic rationality, technological advancement, environmental friendliness, and resource conservation.

1.0.2 This code is applicable to the design of RCC arch dams on rock foundation for the construction, renovation and extension of hydropower projects.

1.0.3 In addition to this code, the design of RCC arch dams shall comply with other current relevant standards of China.

2 Terms

2.0.1 roller-compacted concrete (RCC)

concrete of no-slump consistency in its unhardened state that is spread and compacted in layers by vibratory rollers

2.0.2 grout-enriched vibrated RCC

concrete compacted by immersion vibrator after adding a certain proportion of grout in the RCC mixture

2.0.3 RCC arch dam

arch dam constructed of RCC

2.0.4 transverse joint

straight joint formed by providing an inducing system or pouring concrete by blocks, which is set at a transverse or approximate transverse section in the dam to disable the crack resistance of this section, with the function of releasing the temperature stresses during the construction period

2.0.5 induced joint

partial joint formed by providing an inducing system, which is set at a transverse or approximate transverse section in the dam to reduce the crack resistance of this section, with the function of releasing the temperature stresses during the construction period

2.0.6 RCC layer joint

bonding surface between upper and lower RCC layers

2.0.7 mineral admixture

active or inactive mineral materials added to concrete mixture to improve the performance of concrete and reduce cement consumption

2.0.8 layer placement duration

time period from the completion of roller compaction of an RCC layer to the compaction completion of the next layer

3 Basic Requirements

3.0.1 RCC arch dams may be classified by height into low dams with a height below 30 m, medium dams with a height of 30 m to 70 m, and high dams with a height greater than 70 m.

3.0.2 RCC arch dams may be classified by the base thickness/height (B/H) ratio into thin arch dams with a B/H ratio less than 0.20, medium-thick arch dams with a B/H ratio of 0.20 to 0.35, and thick arch dams or gravity arch dams with a B/H ratio greater than 0.35.

3.0.3 In the design of RCC arch dams, the basic data on meterology, hydrology, sedimentation, topography, geology, seismicity, construction materials, ecology, and environment shall be collected. and the construction conditions shall be studied to ensure the safety, economy and serviceablity of the project.

3.0.4 The design of RCC arch dams shall:

1. Optimize arch dam layout, select dam shape and basic geometry to meet the safety and function requirements and to facilitate the rapid construction of RCC.

2. Select raw materials of dam concrete, determine the strength classes and related design parameters, and determine the concrete zoning.

3. Arrange the joints and conduct the temperature control design to improve the performance of crack prevention of RCC.

4. Propose the construction requirements for concrete placement and joint grouting to ensure structural safety.

3.0.5 The hydraulic and foundation treatment design of RCC arch dams shall comply with the current sector standard DL/T 5346, *Design Specification for Concrete Arch Dams*.

3.0.6 For the design of RCC arch dams with a height greater than 150 m or of particular importance, special issues shall be studied in addition to complying with this code.

4 Layout of RCC Arch Dam

4.1 General Requirements

4.1.1 An RCC arch dam should be built at a rock-foundation site in a narrow valley, with favorable geological conditions, raw material sources and construction conditions for RCC.

4.1.2 The axis selection of the RCC arch dam shall meet the following requirements:

1. Rock masses at both abutments shall be relatively intact to satisfy the stability requirements.
2. The flood discharge and energy dissipation conditions shall be favorable.
3. High cut slopes should be avoided.

4.1.3 The layout of an RCC arch dam shall be determined through techno-economic comparison according to the topographical, geological, and hydrological conditions at the dam site on the premise of fulfilling the functional requirements and the intended purpose.

4.1.4 The layout of an RCC arch dam and other structures, such as the release structures, headrace and powerhouse structures, shall be designed as a whole according to their importance, type, construction conditions, and operation and maintenance requirements.

4.1.5 The structural configuration of an RCC arch dam should facilitate the rapid construction and wider use of RCC.

4.2 Dam Shape Design

4.2.1 The dam shape for an RCC arch dam shall be determined through comparison of alternatives, considering the factors such as the canyon shape, geological conditions, layout and type of structures in the dam, dam stresses and abutment stability, and construction conditions.

4.2.2 The dam shape design shall meet the following requirements:

1. The dam stresses shall satisfy the control criteria and foundation bearing capacity requirements, and distribute reasonably.
2. The maximum central angle of the arches should be 75° to 110°.
3. The overhang ratio of the cantilevers should not exceed 0.25.
4. The impact of the release structures on the integrity of the arch dam

shall be considered.

    5   On the premise of fulfilling the stress control criteria, the intersection angle between the thrust force and the assumed line of competent rock should be increased. The angle between the intrados tangent at the arch abutment and the assumed line of competent rock should not be less than 30°.

    6   The requirements for rapid RCC construction, construction techniques and construction equipment operation should be met.

4.2.3   For the shape design of a high RCC arch dam, the sensitivity analysis shall be conducted for the modulus of elasticity of RCC, comprehensive deformation modulus of the foundation, and thermal actions.

4.2.4   Gravity blocks or thrust blocks may be provided at one or both banks in any of the following cases:

    1   Wide upper valley.

    2   Locally unsound rock, which is unsuitable to be the foundation of an arch dam.

    3   Unable to meet the abutment stability requirements.

4.2.5   If the foundation of an RCC arch dam has local geological defects, concrete pads may be provided to fulfill the stress and stability requirements.

4.2.6   If the symmetricity in shape or geological condition of the valley at the arch dam site is relatively poor, appropriate measures shall be taken to improve the dam stresses and stability conditions.

4.3   Structures in Dam Body

4.3.1   The crest spillway should be given priority as the RCC arch dam release structures. If high-level, low-level or bottom outlets are needed, the layers and number of outlets in the dam should be minimized as appropriate.

4.3.2   When the water release, diversion and intake outlets are set in the dam, the size, number, location and shape of the outlets shall be determined based on factors such as the function and safety requirements, dam stresses, and the relation with other structures.

4.3.3   The arrangement of the access, grouting and monitoring galleries should be simple and reasonable to minimize the impacts on dam stresses and the interference with RCC construction, and facilitate flood control during the construction period.

4.3.4 If the powerhouse is located at the dam toe, the layout of the pressure conduits and the powerhouse shall be determined by techno-economic comparison based on the dam stresses, the layout of water release and energy dissipation structures, construction and operation conditions, etc.

5 RCC of Dam

5.1 General Requirements

5.1.1 The cement, aggregate, water, mineral admixtures and chemical admixtures used in RCC shall comply with the current standards of China.

5.1.2 Moderate-heat cement or lower heat-of-hydration ordinary Portland cement should be used in RCC. Special requirements for chemical components, mineral composition, heat of hydration, fineness, etc. should be proposed according to particular needs of the project.

5.1.3 Active mineral admixtures such as fly ash, granulated blast-furnace slag, and volcanic ash should be used in RCC.

5.1.4 The aggregate sources shall be determined through techno-economic comparison considering factors such as the distribution and reserves of construction materials, source material quality, mining conditions, mixture proportioning study and construction quality control.

5.1.5 For the aggregate used in dam and appurtenant structures, alkali-aggregate reactivity test shall be conducted. The aggregates with alkali-carbonate reactivity shall not be used. For the use of the aggregate with alkali-silicate reactivity, mitigation measures shall be adopted after demonstration.

5.1.6 Chemical admixtures such as water-reducing retarder and air-entraining agents should be used in RCC, which can reduce water consumption and meet the requirements for roller compaction, set retarding and durability.

5.1.7 The RCC shall meet the requirements for strength, permeability, durability and low heat of hydration.

5.1.8 When the environmental water exhibits corrosivity, mitigation measures such as suitable cement, mineral admixtures, chemical admixtures and mixture proportion shall be adopted and verified through tests.

5.2 Strength Class of RCC

5.2.1 The strength class of RCC shall be determined by the compressive strength, with a dependability of 80 %, of the 150 mm cubic specimens made and cured using the standard method, which is measured by the standard test method at the design age, notated as C_d strength (MPa), where d is the design age of the concrete. The design age should be 90 d, and may be 180 d after demonstration.

5.2.2 The strength class of RCC shall not be inferior to $C_{90}15$.

5.3 Physical and Mechanical Properties of RCC

5.3.1 If the compressive strength at the non-design age is used for dam RCC, the rate of compressive strength gain with age should be determined by tests. For medium and low dams, the compressive strength may be taken with reference to that of similar projects in the absence of test data.

5.3.2 The characteristic value of the RCC's tensile strength may be taken as 0.08 to 0.10 times the characteristic value of its compressive strength.

5.3.3 The unit weight of RCC should be determined by tests. In the absence of test data, it may be taken as per the current sector standard DL 5077, *Specifications for Load Design of Hydraulic Structures*.

5.3.4 The elastic modulus, Poisson's ratio, linear expansion coefficient, and ultimate tensile strain should be determined by tests. In the absence of test data, these parameters may be taken with reference to those of similar projects.

5.4 Seepage Resistance and Durability Performance of RCC

5.4.1 The seepage resistance of RCC shall be determined by the working head, and may be taken as per Table 5.4.1.

Table 5.4.1 Seepage resistance of RCC

No.	Position	Working head H (m)	Seepage resistance
1	Interior of dam	-	W4
2	Upstream impervious zone of dam	$H < 50$	W6
		$50 \leq H < 100$	W6 - W8
		$100 \leq H < 150$	W8 - W10

5.4.2 The impervious zone of an RCC arch dam should adopt RCC with a maximum aggregate grain size of 40 mm. For the impervious zone, the thickness should be 1/20 to 1/15 of the upstream working head, and the minimum thickness shall meet the RCC construction requirements. For low RCC arch dams, the RCC with a maximum aggregate grain size of 80 mm may be used for the impervious zone.

5.4.3 The seepage resistance of RCC shall be obtained by testing specimens at 90 d age, and the test method shall comply with the current sector standard DL/T 5433, *Test Code for Hydraulic Roller Compacted Concrete*.

5.4.4 The frost-resistance of RCC shall be determined comprehensively according to the factors such as the climate zone, freeze-thaw cycles, local microclimate conditions near the dam surface, saturation degree, and importance and maintenance conditions of the structural members, and it shall comply with the current sector standard NB/T 35024, *Design Code for Hydraulic Structures Against Ice and Freezing Action*.

5.4.5 For RCC arch dams in a severe cold region, permanent thermal insulation coats or boards shall be applied on the upstream surface in the drawdown zone and the downstream surface.

5.4.6 According to the durability requirements of dam concrete, the mixture proportion of RCC shall comply with the current sector standard DL/T 5241, *Technical Specifications for Durability of Hydraulic Concrete*.

6 Stress and Stability Analysis of Arch Dams

6.0.1 The dam stresses and abutment stability shall be calculated for an RCC arch dam. When necessary, appropriate engineering measures shall be taken to ensure dam safety.

6.0.2 The stress analysis, abutment stability analysis, overall stability analysis, and the geomechanical model tests of an RCC arch dam shall comply with the current sector standard DL/T 5346, *Design Specification for Concrete Arch Dams*.

6.0.3 The stress and stability analysis of an RCC arch dam under seismic action shall comply with the current sector standard NB 35047, *Code for Seismic Design of Hydraulic Structures of Hydropower Project*.

6.0.4 The trial load method shall be used for the stress analysis of an RCC arch dam. For high dams or arch dams with large openings or large appurtenant structures, the finite element method should also be used.

6.0.5 In the stress calculation of an RCC arch dam, factors such as the configuration of the dam, loads, foundation, geological defects, construction process, and impoundment process shall be considered to ensure that the arch dam satisfies the stress requirements under various operating conditions and has relatively good adaptability.

6.0.6 The closure temperature of an RCC arch dam shall be determined comprehensively according to the stable temperature field, air temperature conditions, temperature control conditions, and dam stress calculation results for the construction and operation periods. If transverse and induced joints are not provided or the joint spacing is too large, the highest temperature field of each concrete pouring zone before the design age should be selected as the closure temperature, or the residual temperature stresses during the construction period should be taken into consideration in the dam stress calculation.

6.0.7 For high RCC arch dams or RCC arch dams with complicated geological conditions, nonlinear analysis of the dam-foundation system shall be conducted by the finite element method, and the geomechanical model tests shall be performed when necessary.

7 Dam Detailing

7.1 Dam Crest Layout

7.1.1 The width of the dam crest shall be determined by the dam stress calculation, layout of the dam crest and construction requirements, and should not be less than 5 m. A cantilever structure may be provided at the upstream or downstream side of the dam crest if there are traffic or other layout requirements.

7.1.2 The determination of dam crest elevation, and the layout of structures such as parapet wall, access bridge, roadway, walkway, and cable trench shall comply with the current sector standard DL/T 5346, *Design Specification for Concrete Arch Dams*.

7.2 Zoning Design

7.2.1 The concrete zoning of an RCC arch dam (see Figure 7.2.1) shall be designed according to the position and working environment. The performance requirements for dam concrete zoning shall be in accordance with Table 7.2.1.

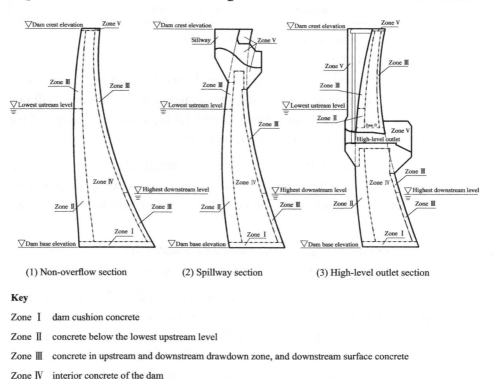

(1) Non-overflow section   (2) Spillway section   (3) High-level outlet section

Key

Zone I   dam cushion concrete

Zone II   concrete below the lowest upstream level

Zone III   concrete in upstream and downstream drawdown zone, and downstream surface concrete

Zone IV   interior concrete of the dam

Zone V   conventional concrete of the dam

Figure 7.2.1 Concrete zoning of the dam

Table 7.2.1 Performance requirements for dam concrete zoning

Zone	Strength	Seepage resistance	Frost resistance	Corrosion resistance	Low heat	Maximum water-cementitious materials ratio	Main factors for zoning
I	++	++	+	++	++	+	Seepage resistance and crack prevention
II	++	++	+	++	++	+	Seepage resistance and crack prevention
III	++	++	++	++	++	+	Frost resistance and crack prevention
IV	++	+	+	+	++	+	Crack prevention
V	The characteristics of conventional concrete are determined as required.						

NOTE ++ denotes the main control factors for selecting the concrete classes for various zones, and + denotes the factors that need to be required.

7.2.2 RCC of different strength classes may be used for different elevations or positions. For any two adjacent zones, the strength class difference shall not exceed 5 MPa for RCC and should not exceed 10 MPa for the strength class difference between RCC and conventional concrete.

7.2.3 The zone width of RCC with different strength classes or aggregate gradations should be determined by the requirements for dam stresses, durability, detailing, and construction conditions, and should not be less than 3 m.

7.2.4 Grout-enriched vibrated RCC zones with a width of 0.5 m to 1.0 m shall be provided at both upstream and downstream surfaces of an RCC arch dam.

7.2.5 Considering the foundation surface roughness in the riverbed, a cushion of conventional concrete or a levelling layer of grout-enriched vibrated RCC or mortar should be placed prior to the dam RCC construction. The cushion/layer thickness may be determined according to the foundation surface roughness, temperature control, and foundation consolidation grouting requirements.

7.2.6 Grout-enriched vibrated RCC zones should be adopted at both abutments, and their width should be 1.0 m.

7.3 Contraction Joints

7.3.1 Transverse joints, induced joints or other joints may be adopted for an RCC arch dam.

7.3.2 Transverse joints or induced joints shall be set according to temperature control and crack prevention requirements, geological conditions of the dam foundation, dam structure layout and construction conditions. Longitudinal joints should not be set.

7.3.3 Keys shall be provided for transverse joints, and waterstops and grouting systems shall be set in transverse joints. The key geometry may be trapezoidal, circular arc or spherical.

7.3.4 The waterstops and grouting systems shall be set in induced joints. The induced joint area should not be less than 25 % of the whole section area.

7.3.5 The transverse or induced joint may be a vertical joint in the radial or approximately radial direction, or a twisted joint in the radial direction. The included angle between the joint face and the dam foundation surface should not be less than 60º.

7.3.6 The spacing between induced joints or transverse joints shall be determined according to temperature control and crack prevention calculation or engineering analogies, and should be 30 m to 50 m, taking into account such factors as modulus and uniformity of dam foundation rock mass and abutment slope.

7.3.7 For the contacting joints between the RCC of the dam and the surrounding concrete of the outlet not poured at the same time, keys and joint-crossing rebars shall be set, and a grouting system may be set as required.

7.4 Joint Grouting

7.4.1 Transverse joints and induced joints of dam shall be grouted. The dam shall not be used for long-term water retaining until the cement grout reaches the design strength.

7.4.2 When the uncompleted dam is temporarily used for water retaining during floods in the construction period or the staged impoundment period and the joint grouting of some arches has not been completed, the safety of arch dam shall be analyzed and demonstrated.

7.4.3 The joint grouting system should be reusable.

7.4.4 The joint grouting shall meet the following requirements:

    1 The concrete temperature of the grouting zone shall drop to the design value.

2 The thickness of the weighted cover above the grouting zone should not be less than 6 m.

3 The concrete temperature difference between the weighted cover and the grouting zone should not exceed 3 °C.

4 The age of the concrete at both sides of a joint should not be less than 90 d, and the age of weighted cover concrete should not be less than 28 d.

5 For the joint with an opening greater than 0.5 mm, cement may be used as grouting material. Otherwise, superfine cement or chemical material should be used.

7.4.5 The joint grouting pressure should be 0.3 MPa to 0.6 MPa, and the grouting pressure at the top zone may be reduced properly.

7.4.6 The area of the grouting zone for the transverse joint and induced joint should be 300 m^2 to 450 m^2, and the height of the grouting zone should not be greater than 15 m.

7.5 Galleries and Adits

7.5.1 Grouting, drainage, monitoring and access galleries may be set inside the dam. The structural arrangement of galleries shall be determined according to their functions and impacts on the dam layout and structures.

7.5.2 The galleries shall avoid the tensile stress area of the dam. The clear distance between a gallery and an opening shall meet the dam stress and local structural requirements. The distance between the upstream wall of a longitudinal gallery and the upstream face of dam should not be less than 0.05 to 0.10 times the working head, and not less than 3 m.

7.5.3 The structural arrangement of galleries in a dam shall consider factors such as access and ventilation conditions. Sufficient lighting facilities should be provided inside the galleries. All equipment and electric lines shall be well insulated and easy for maintenance.

7.5.4 The galleries at different elevations should be connected by access bridge behind the dam, stairway, elevator, or shaft, and the layout should facilitate construction and access during the operation period.

7.5.5 When the floor of the drainage gallery is above the highest downstream water level, gravity drainage may be adopted; otherwise, a pump-drainage system shall be set.

7.5.6 Protective measures shall be taken for the portals of galleries prone to flooding.

7.5.7 The galleries, drainage equipment and access facilities shall meet the fire safety requirements.

7.5.8 The impacts of dam deformation shall be considered in the structural arrangement of appurtenant structures such as the shaft inside the dam and elevator shaft.

7.6 Waterstops and Drains

7.6.1 Waterstops shall be provided at the waterward side of the joints and contacting surfaces between the dam and steep abutment. Two or more waterstops shall be set in the case of a relatively high working head, and the spacing between waterstops should be 0.3 m to 0.5 m. The material and structural type of waterstops shall comply with the current sector standards DL/T 949, *Standard for Joint Plastic Sealant of Hydraulic Structure,* and DL/T 5215, *Specification for Waterstop of Hydraulic Structure.*

7.6.2 Drainage facilities may be set behind the impervious zone of the dam as required.

8 Crack Prevention and Temperature Control

8.1 General Requirements

8.1.1 The temperature control and crack prevention design of an RCC arch dam shall be conducted by comprehensively considering the thermal properties and crack prevention characteristics of concrete, structure characteristics, constraint conditions, and construction characteristics such as placement in full face, thin layer, and continuous roller compacting.

8.1.2 In the temperature control and crack prevention design, attention shall be paid to improving the crack resistance of RCC, and proper raw materials and mixture proportions of concrete shall be adopted.

8.1.3 For high and medium-high RCC arch dams, the temperature field and temperature stresses shall be simulated and analyzed using the finite element method, and the temperature control criteria and requirements, joint design, pouring schemes, and other crack prevention measures shall be proposed. For low RCC arch dams, the joint design and the temperature control and crack prevention design may refer to the experience of similar projects.

8.1.4 For high RCC arch dams, the mechanical, thermal and deformation properties of RCC shall be obtained by tests. For medium-high and low RCC arch dams, necessary tests may be performed if required, and the values may be taken by referring to similar projects in the absence of test data.

8.1.5 The following data shall be collected for the temperature control and crack prevention design:

1. Mean annual air temperature and intra-annual variation, mean monthly air temperature and mean ten-day air temperature at the dam site.

2. Extreme air temperature, and amplitude, duration and corresponding frequency of sudden air temperature drop at the dam site.

3. Mean annual water temperature at the dam site.

4. Earth temperature of the dam foundation.

5. Solar radiation and wind speed at the dam site.

6. Reservoir water temperatures of similar projects.

8.1.6 Attention shall be paid to the water overtopping and retaining of the dam body during the construction period in the temperature control and crack prevention design.

8.1.7 The temperature control and crack prevention design of an RCC arch dam shall comply with the current sector standard NB/T 35092, *Design Code for Temperature Control of Concrete Dam*.

8.2 Temperature Control Standard

8.2.1 When the ultimate tensile strain of RCC at 90 d age is not less than 0.75×10^{-4}, the coefficient of linear thermal expansion for the aggregate used in concrete is not much different from 1.0×10^{-5} /°C, the deformation modulus of the rock foundation is close to the elastic modulus of concrete, and the concrete pouring is uniform with a short interval, the allowable temperature difference of foundation may be adopted according to Table 8.2.1.

Table 8.2.1 Allowable temperature difference of foundation ΔT (°C)

Height above foundation surface h	Length of long side of concrete pouring block L		
	Below 30 m	30 m - 70 m	Over 70 m
$(0 - 0.2) L$	18.0 - 15.5	15.5 - 12.0	12.0 - 10.0
$(0.2 - 0.4) L$	19.0 - 16.5	16.5 - 14.5	14.5 - 12.0

8.2.2 The allowable temperature difference of concrete shall be analyzed and demonstrated for the following cases:

1 Concrete block with long pouring interval or overtopping in the foundation restraint area.

2 Significant difference between the deformation modulus of rock foundation and elastic modulus of concrete.

3 Dental concrete or replacement concrete at the dam foundation.

4 Significant autogenous volume change of RCC by tests or measurement.

8.2.3 For the concrete in the constraint area on a steep slope, the allowable temperature difference should take the lower value in Table 8.2.1 of this code or 1 °C to 2 °C lower than that of the concrete blocks on a gentle slope.

8.2.4 The allowable temperature difference between the interior and exterior of dam concrete should be 15 °C to 20 °C.

8.2.5 The allowable temperature difference between the newly placed concrete and the underlying lift should be 15 °C to 20 °C. When the long side length of the concrete block is greater than 40 m, the constraint stresses should be determined through calculation.

8.2.6 The maximum allowable temperature of dam concrete shall satisfy the requirements for the allowable temperature differences of foundation concrete and between the interior and exterior of the dam. For the condition of the long interval, the allowable temperature difference requirement for new-placed concrete and underlying lift concrete shall also be met.

8.3 Temperature Control Measures

8.3.1 The temperature of bulk cement should not exceed 65 °C before unloading into silos.

8.3.2 Aggregate silo shall have sufficient storage volume, the height of the stockpile should not be less than 6 m, and measures such as sunshade and rainproof shed and aggregate-feeding underground gallery shall be taken to reduce the aggregate temperature.

8.3.3 Measures such as precooling aggregates and mixing with chilled water should be taken to control the concrete temperature at the mixer outlet. Air cooling, immersion cooling, chilled water spray, etc., may be adopted as aggregate precooling measures.

8.3.4 For the hauling vehicles or equipment of RCC, measures such as sunshade and rainproof shed should be taken.

8.3.5 RCC hauling, spreading and compacting shall be expedited to control the pouring temperature.

8.3.6 The base concrete should be placed in the low-temperature season or low-temperature period.

8.3.7 Measures such as mist spray, thermal insulation and moisture preservation at the concrete placing field shall be taken during the concrete construction in hot season. Measures such as thermal insulation and frequent curing shall be taken for regions with significant daily temperature differences.

8.3.8 Pipe cooling should be adopted for high and medium-high RCC arch dams, and the layout of cooling pipes, cooling time, water temperature and flow rate shall be determined with the aid of 3D finite element simulation and with reference to similar projects.

8.3.9 Curing shall start upon the final set of RCC, and should last for no less than 28 d.

8.3.10 For dam surface and lift surface subject to long halting period in winter, cold wave, or water passing through the dam notch during construction, the temperature control and crack prevention measures shall be specially studied. The concrete age should not be less than 14 d when the water passes the lift surface.

9 Safety Monitoring Design

9.1 General Requirements

9.1.1 Instrumentation shall be provided for an RCC arch dam according to the grade of structure, dam height, structural characteristics, topographical and geological conditions, and environment, to monitor the dam performance and safety during the construction, initial impoundment and operation period, to guide the construction and operation, and to provide feedback to design.

9.1.2 The monitoring of an RCC arch dam shall cover the dam, dam foundation, abutments, the conveying and discharging structures or equipment directly related to the dam safety, and the nearby bank slope with great influence on the dam safety.

9.1.3 In addition to this code, the monitoring design of an RCC arch dam shall comply with the current sector standards DL/T 5178, *Technical Specification for Concrete Dam Safety Monitoring;* DL/T 5416, *Specification of Strong Motion Safety Monitoring for Hydraulic Structures;* and DL/T 5211, *Technical Specification for Dam Safety Monitoring Automation.*

9.2 Monitoring Items and Arrangement

9.2.1 The safety monitoring shall include patrol inspection and instrumentation.

9.2.2 The positions and items of patrol inspection shall cover the following:

1. The patrol inspection shall cover the dam crest, upstream and downstream dam surfaces, galleries, structure joints, dam foundation, abutments, etc.

2. The items of patrol inspection shall include cracking, spalling, dislocation, crushing and leakage, etc.

9.2.3 The instrumentation of an RCC arch dam shall be provided for monitoring of deformation, stress and strain, seepage and seepage pressure, and environmental quantities.

9.2.4 The dam seepage pressure monitoring shall take the RCC layer joints or lift joints as representative monitoring sections. Piezometers should be embedded in the upstream impervious zone or close to the interface between the impervious zone and the interior RCC at the monitoring section.

9.2.5 The dam temperature monitoring shall select representative vertical and horizontal monitoring sections. The number of vertical monitoring sections should not be less than 2, and the number of horizontal monitoring sections should be 3 to 7, considering the dam height.

9.2.6 Joint meters shall be provided for monitoring the transverse joints, induced joints and contact face between the dam and foundation. The joint meters of the transverse and induced joints in the dam should be set in the mid-part of the grouting zone. For monitoring the hazardous cracks during the construction, joint meters or crack meters shall be provided at the representative cracks.

9.3 Requirements for Instrument Installation

9.3.1 The pitting-and-embedding method should be adopted to install the interior instruments and cables in the RCC dam. For instruments without and with the directional requirement, the pit depth shall be such that the backfilled protective layer is not less than 20 cm and 50 cm after instrument installation, respectively.

9.3.2 The backfilling concrete in the pit shall be manually placed and compacted in layers to guarantee good bonding with the surrounding concrete, and the aggregate with a grain size greater than 40 mm shall be removed from the backfilling concrete.

9.3.3 Extra length shall be reserved for the monitoring cables near the embedding position.

9.3.4 The installed instruments shall be marked and protected. No construction machinery is allowed to pass the backfilled concrete of the embedded area until the initial set of the backfilled concrete or the spread of the upper RCC layer.

10 Construction Requirements

10.0.1 Material sourcing principle shall be defined and the requirements for quarrying plan shall be proposed, according to demonstration result in previous stages.

10.0.2 The critical thermal and mechanical properties of concrete shall be specified, considering construction conditions and demonstration results in previous stages.

10.0.3 The VC value at mixer outlets and lifts, allowable layer placement duration, and measures when exceeding allowable layer placement duration shall be proposed, considering construction condition and experience of similar projects.

10.0.4 The test fill requirements shall be proposed for high arch dams and important medium-high arch dams. Test fill and coring can be used to determine the construction mixture proportion, verify the quality control parameters such as VC value at lifts and layer placement duration, and define the construction parameters.

10.0.5 RCC mixing, hauling and placing capacity shall match the placing intensity within the allowable layer placement duration. The equipment and means for hauling and spreading which can reduce aggregate segregation shall be adopted.

10.0.6 For high RCC dams, the requirements for digital construction management should be proposed on raw materials, mixing, hauling, roller compacting, temperature control, and crack prevention, considering construction condition.

10.0.7 Safety monitoring and data compiling requirements shall be proposed according to the project characteristics and needs for temperature control, crack prevention, slope safety, flood control safety, etc.

10.0.8 Technical requirements for the treatment of dam concrete defects such as appearance defect, crack, and leakage shall be proposed according to the construction conditions and experiences of similar projects. For deep or through cracks and large-scale leakage, specific treatment plans shall be proposed according to the position, scale, shape and developing trend. Seepage control strengthened treatment shall be conducted for cracks at upstream dam surface, and structural integrity shall be recovered after treatment of through cracks.

10.0.9 The timing of joint grouting shall consider the dam body temperature, the opening of transverse and induced joints, flood control requirement, construction schedule, impoundment plan, etc.

10.0.10 Flood control schemes during construction period, as well as corresponding physical progress and engineering measures, shall be proposed, considering the construction conditions and previous demonstration results.

11 Initial Impoundment, Operation and Maintenance

11.0.1 Before the initial impoundment, the requirements for physical progress, pool level rise rate, items and frequency of patrol inspection and instrumentation, and the flood control scheme during the initial impoundment shall be proposed considering the specific project conditions.

11.0.2 The physical progress shall meet the following requirements before initial impoundment:

1. The finished elevation of dam is higher than the flood level of the corresponding flood standard, with a freeboard over 1 m and a concrete age of more than 28 d.

2. The dam foundation consolidation grouting, curtain grouting, contact grouting and joint grouting below the pool level are completed.

3. The reinforcement for the abutments is conducted.

4. The construction defects are repaired.

5. The exploration adits and boreholes in the dam foundation and abutments are treated, and the dam cooling pipes are plugged.

6. The water release and intake structures required for initial impoundment are completed, and the corresponding gates and electromechanical equipment are installed and tested.

7. The downstream energy dissipation, scour-resistant structures and atomization zone treatment are finished.

8. Safety evaluation and treatment of unfavorable geological bodies such as deposits and landslides in the reservoir area are completed.

9. Instrumentation for the dam body, abutments and near-dam slopes is completed, and the initial observation data are obtained.

10. Power supply, lighting, communication and hydrological telemetry and forecast system are up and running, and the emergency power supply is ready.

11.0.3 During initial impoundment, the pool level rise rate below the high-level outlets need not be controlled, but should be limited to 2 m/d to 5 m/d above the high-level outlets. The closer the pool level approaches the normal pool level, the lower the rise rate shall be. One or more observation periods for impoundment should be specified for high pool levels, which may be 5 d to 7 d.

The pool level can continue to rise provided that the dam is deemed safe based on comprehensive analysis.

11.0.4 The pool level rise and drop rates during the normal operating period shall be proposed according to the dam safety requirements and reservoir operation and flood regulation conditions, considering the dam structure and foundation condition and stability condition of unfavorable geological bodies in the reservoir.

11.0.5 The schemes and requirements for emergency drawdown in case of landslide, earthquake and other hazards shall be proposed considering the project-specific conditions.

11.0.6 The routine instrumentation items and frequency, and position and items of patrol inspection shall be clearly defined according to the characteristics and safety requirements of the project, and more frequent inspections shall be conducted before, during and after the flood season, and at the end of the winter each year. The identified defects and potential hazards shall be remedied timely.

11.0.7 The water release structures and relevant equipment and devices shall be inspected and maintained before flood season to ensure the discharge capacity and normal operating conditions. The release structures and gates shall be regulated and operated according to the reservoir operation and power dispatch rules specific for this project.

11.0.8 The dam shall be checked in the event of a big flood, felt earthquake or extreme weather. Emergency preparedness plan shall be activated timely in case of natural disasters such as severe flood, destructive earthquake, or other catastrophic events. After the occurrence, the dam safety shall be checked and evaluated, and the flood standards and earthquake ground motion parameters shall be reviewed and verified.

11.0.9 When significant variation is observed in the verification results for flood, flood regulation and seismic parameters, or when the flood control or seismic standard is raised above design, a flood or earthquake safety check shall be conducted, and dam safety shall be evaluated.

11.0.10 The special treatment for major engineering defects and hazards arising from dam operation shall be designed and be implemented timely.

Explanation of Wording in This Code

1. Words used for different degrees of strictness are explained as follows in order to mark the differences in executing the requirements in this code:

   1) Words denoting a very strict or mandatory requirement:

      "Must" is used for affirmation, "must not" for negation.

   2) Words denoting a strict requirement under normal conditions:

      "Shall" is used for affirmation, "shall not" for negation.

   3) Words denoting a permission of a slight choice or an indication of the most suitable choice when conditions permit:

      "Should" is used for affirmation, "should not" for negation.

   4) "May" is used to express the option available, sometimes with the conditional permit.

2. "Shall meet the requirements of…" or "shall comply with…" is used in this code to indicate that it is necessary to comply with the requirements stipulated in other relative standards and codes.

List of Quoted Standards

NB/T 35024,	*Design Code for Hydraulic Structures Against Ice and Freezing Action*
NB 35047,	*Code for Seismic Design of Hydraulic Structures of Hydropower Project*
NB/T 35092,	*Design Code for Temperature Control of Concrete Dam*
DL/T 949,	*Standard for Joint Plastic Sealant of Hydraulic Structure*
DL 5077,	*Specifications for Load Design of Hydraulic Structures*
DL/T 5178,	*Technical Specification for Concrete Dam Safety Monitoring*
DL/T 5211,	*Technical Specification for Dam Safety Monitoring Automation*
DL/T 5215,	*Specification for Waterstop of Hydraulic Structure*
DL/T 5241,	*Technical Specifications for Durability of Hydraulic Concrete*
DL/T 5433,	*Test Code for Hydraulic Roller Compacted Concrete*
DL/T 5346,	*Design Specification for Concrete Arch Dams*
DL/T 5416,	*Specification of Strong Motion Safety Monitoring for Hydraulic Structures*